# James Hutton

## The Founder of Modern Geology

Donald B. McIntyre and Alan McKirdy

EDINBURGH: THE STATIONERY OFFICE

First published 1997

The Stationery Office Limited
South Gyle Crescent, Edinburgh EH12 9EB



British Library Cataloguing in Publication Data

A catalogue record for this book is available from the British Library

Photographs not individually credited are by Donald B. McIntyre.

ISBN 0 11 495812 2

*Cover illustrations:*

*Front:* *Portrait of James Hutton by Sir Henry Raeburn. This important portrait was formerly considered to have been painted before the artist went to Rome in 1784. Dr David Mackie has, however, shown that it is much later; he dates it as c. 1790 (Ph.D. thesis, Edinburgh). Scottish National Portrait Gallery.*
*Veins of red granite in a dark older rock provide the background. These veins were Hutton's evidence that granite was an intrusive igneous rock and not the oldest of all rocks.*

*Back:* *The Earth from space. Hutton had a holistic view of the Earth as a habitable planet (NASA).*

# Contents

*John Clerk of Eldin, by Archibald Skirving, c. 1800 (from a private collection). Photograph by Stephen Kearny.*

This book is dedicated to the memory of John Clerk of Eldin

*The interest he took in studying the surface no less than the interior of the earth; his extensive information in most branches of natural history; a mind of great resource, and great readiness of invention; made him, to Dr Hutton, an invaluable friend and coadjutor. It cannot be doubted, that, in many parts, the system of the latter has had great obligations to the ingenuity of the former, though the unreserved intercourse of friendship, and the adjustments produced by mutual suggestion, might render those parts undistinguishable even by the authors themselves.*

**John Playfair**

# THE DYNAMIC EARTH PROJECT AND THE NEXT MILLENNIUM

250 years ago Edinburgh was a focus of intellectual ferment. The Enlightenment fired concern for an understanding of science of the natural world and its social, political and economic organisation. Hutton's *Theory of the Earth* not only provided a comprehensive view of the planet and its processes but also altered the perception of time. However, these great thoughts impinged little on the artisans of Edinburgh who inhabited the closes and wynds of the Royal Mile. The next Millennium will see a new global Enlightenment. The communications revolution is extending an awareness of science but there is also an increased responsibility on scientists to make the results of their work understandable and accessible to the general public.

In Edinburgh, the ***Dynamic Earth*** exhibition will be housed in a landmark building, the William Younger Centre, located within half a mil of the house where Hutton wrote the *Theory of the Earth* and set against the backdrop of Salisbury Craigs. It will be owned by the ***Dynamic Earth Charitable Trust*** whose objectives include "promoting awareness and understanding of . . . the Dynamic Earth through the establishment of a permanent exhibition and educational centre for the public benefit". The exhibition will present a holistic view of the planet and generate that sense of excitement felt by Hutton and his colleagues when first they looked into ". . . the abyss of time." It will demonstrate that the Earth is an active and dynamic system on many scales, and show how landscapes are formed by a complex interaction between the geosphere, atmosphere, hydrosphere and biosphere and that these processes continue today.

The challenge of Hutton's concept that "the present is the key to the past" lies in what it implies about the future. The presentation within the Dynamic Earth will help visitors make up their own mind about what the future may hold.

The Dynamic Earth Charitable Trust is a company limited by guarantee
Registered in Scotland, Number 138695 and by the Inland Revenue as a Charity, Number Sco 20363

# FOREWORD

JAMES HUTTON was a genius, a star of the Scottish Enlightenment, that constellation of friends and kin that gave the world modern philosophy, modern economics and much of modern science. In that great age of the eighteenth-century intellect, none made a more original contribution than Hutton to our understanding of the planet. He simply gave us a true sense of time. It is hard to imagine now that until a few generations ago it was assumed that the Earth was only 6,000 years old. It was the investigations of this enormously vigorous, brilliant and sociable man, travelling the length of Scotland and England to scan the strata and poke among the rocks, that began to show how very ancient the Earth really was. In doing so, Hutton laid the foundations of geological science.

In this excellent and highly enjoyable book, Donald McIntyre and Alan McKirdy make the story accessible. They convey, so vividly, the excitement that Hutton himself had about his discoveries. They describe the history as well as the science, showing Hutton's friendships with other great figures in the Enlightenment, and the open character of a society where a kirk leader like William Robertson felt perfectly free to publicise Hutton's theories, even though they called into question the previous assumptions of the Church. They explain the importance of the scientific discoveries and make us want to go to view the very places, like Glen Tilt and Siccar Point, where Hutton saw, for the first time, that his theories about the age of the Earth must be true.

Geology is a marvellous part of Scotland's natural heritage: no land has more varied rocks and soil, or a more exciting history of their discovery. To read this story of James Hutton is an ideal way to begin to understand it all.

Professor T. C. Smout,
CBE, FRSE, FBA, FSA(Scot),
Historiographer Royal in Scotland

# ACKNOWLEDGEMENTS

DONALD MCINTYRE is pleased to acknowledge his debt to the late Professor S. I. Tomkeieff for the stimulation of his lecture on 19 March 1947, commemorating the 150th anniversary of Hutton's death, and for subsequent encouragement; his gratitude to the late Professors F. J. Turner and Herbert M. Evans for having him as speaker on Hutton at the History of Science Club in Berkeley in 1957; to Pomona College, Claremont, California, for providing a superb library and for supporting his study of Hutton and the Scottish Enlightenment over many years; and particularly to Sir John and Lady Clerk of Penicuik for their friendship, hospitality and interest, and for so generously making available the riches of their family archives.

Donald McIntyre is Emeritus Professor of Geology at Pomona College and holds honorary positions at the Universities of St Andrews and Edinburgh.

Alan McKirdy thanks The Stationery Office for commissioning this book to mark the bicentenary of James Hutton's death. Alan is a graduate of Aberdeen University and works for Scottish Natural Heritage as Head of Advisory Services for Earth Sciences, Habitats and Species.

Sir John Clerk of Penicuik, Bt, has generously given permission to reproduce texts and drawings by John Clerk of Eldin, his father, and his grandfather. Permission has also been given for publication as follows: The Scottish Record Office for letters from the Dunglass Muniments; The British Geological Survey for a cross-section of the Midlothian Syncline; and

The National Aeronautical and Space Administration (NASA) for the photograph of the Earth from space.

We thank Lothian and Edinburgh Enterprise Limited and Scottish Natural Heritage for their financial support.

# Introduction

JAMES HUTTON (1726–97) was a leading figure in the Scottish Enlightenment, that remarkable period in the second half of the eighteenth century associated with, among others, David Hume (philosopher and historian), Adam Smith (economist) and Joseph Black (chemist). Hutton provided unequivocal evidence that the Earth was far older than generally believed. His theory, based on extensive field experience, was confirmed from further observations made specifically to test it. Recognising the vast extent of past time, he saw the possibility of evolution, not only of the physical world, but also of living creatures. Eleven years before Darwin was born, Hutton saw *natural selection* as a 'beautiful contrivance' for adapting animals and plants to their changing environments. His understanding of 'deep time' allowed his successors to discover that, in the words of Nobel laureate George Wald: 'We live in a historical universe, one in which not only living organisms but stars and galaxies are born, mature, grow old and die.'

Despite the far-reaching importance of his methodology and conclusions, Hutton is not as well known as Hume, Adam Smith and Joseph Black. Our purpose is to give Hutton's work the public attention it deserves, using his own words where possible and emphasising that his knowledge was based on extensive observation of natural features and especially of rock outcrops. Hutton's *Theory of the Earth* secured his place as the founder of modern geology.

*The Earth from space. Hutton had a holistic view of the Earth as a habitable planet (NASA).*

Chapter One

# Discovering Deep Time

*Weathered tombstone, Kinfauns, Perthshire. White spots of lichen cover the inscription. The sandstone has split along sedimentary layers. Originally deposited as horizontal beds of sand, the layers have been set vertically in the ground.*

**SCIENTISTS** currently think that the Earth is four and a half thousand million years old. Our present knowledge is the culmination of two centuries of research, started by James Hutton, an Edinburgh man, who demonstrated that the Earth was immensely old.

Living on the Earth, we take it for granted. It is the stage on which our dramas are performed. The Earth seems ageless, and changes seem superficial. We see cycles of crops and, on a longer time-scale, forests grow and are felled, but the Earth beneath us and the hills rising above the valleys seem to endure as a changeless framework. If we think about it at all, we expect the Earth to be essentially the same tomorrow as it was yesterday, or for that matter as it was 'in the beginning'. How can we know what happened long ago when there was no one to make a record?

The Greek historian and traveller, Herodotus (*c.* 484–425 BC) realised that Nile sediments record past events, and concluded that Egypt had formed from slowly accumulating river deposits. Sadly, the lesson was lost, and for nearly two thousand years, the Christian world firmly believed that the Earth was created about 4000 BC. The date came from the Bible, literally interpreted as God's own word, and to question that was heresy.

We think we know about the immediate past from our own experience, though in this we deceive ourselves, for our memories are selective. Much of our knowledge, even of our own time, comes from others – who have their own reasons for informing us. Our memory is a collective one, made up from what we have observed ourselves, what we have been told by parents, teachers and friends, and what we have learned from books and the media. For instance, most of what we know about the Scottish patriot Sir William Wallace comes from Blind Harry, the Minstrel, who collected his information 200 years after the events he describes. Although Blind Harry had access to sources now lost, his business was to flatter the nobles he entertained. The aims of chroniclers were not those of modern historians. Stories, like Robert Bruce finding encouragement in the perseverance of a spider, were sermons told to form good character.

In addition to reading written accounts, we can learn about the past from the remains of physical objects, such as buildings, statues, weapons and other artefacts. Where memory and tangible evidence fail, speculation takes their place. In the past, curiosity was satisfied with myths, and it took a long time for myth to evolve into science. Scholars point out that, for Christians, the birth of Jesus is not the natural outcome of preceding development, but the fulfilment of supernatural prophecy. Consequently, all we need to know about the past can be found in the Old Testament. On that basis, Eusebius (*c.* 260–340) and St Jerome (*c.* 340–420) extended chronology back to Adam. It was this Eusebius who sat at the right hand of Constantine and opened the Council of Nicaea in 325. The influence of his *Chronology* lasted so long that Isaac Newton's final work, *The Chronology of the Ancient Kingdoms Amended*, published posthumously in Hutton's lifetime, was an attempt to revise it.

The date of Creation was not questioned. In Shakespeare's *As You Like It* (1600), Rosalind exclaims: 'The poor world is almost six thousand years old.' Even the great scientist Johannes Kepler wrote: 'My book may wait 100 years for a reader, since God has waited 6,000 years for a witness' (1619). And the *History of the Royal Society* (1667) records: 'They have

*Granite disintegrating into sand, Southern California. Individual crystals in the granite have different physical and chemical properties. Resistant minerals like quartz become sand grains; feldspar decays into clay.*

framed such an Assembly in six years, which was never yet brought about in 6,000.'

Archbishop James Ussher attempted greater precision, beginning his *Annals of the World* (1658) with these words: 'In the beginning God created Heaven and Earth. Which beginning of time fell upon the entrance of the night preceding the twenty-third day of Octob. in the year 4004.' Ussher's date was included in the *English Bible*, and was accepted as part of Scripture itself. Indeed, Darwin had published *The Origin of Species* before discovering to his surprise that the date had been added to the biblical text. In his *Principia*, Newton estimated that a red-hot sphere of iron as big as the Earth would take 50,000 years to cool, but the result was ignored as inconsistent with Scripture.

James Hutton was nine months old when Newton died. Others before Hutton had recognised that wind and weather, running water and waves in the sea, all combine to wear away the land. With no sign of renewal, and available time strictly limited, decay was seen as punishment for human sinfulness. Hutton drastically changed this view when he considered 'the habitable earth with regard to its duration and stability'. He was the first to provide inescapable evidence that the Earth was far older than had been imagined. Today we know Hutton was right – the long accepted figure was a million times too small! We explain his method in the following pages.

A tiny, short-lived creature may conclude that you are inert,

*Cones of scree accumulated from the erosion of rock on the hillside above, Sierra Nevada, California. We see an age sequence, because one of the cones is superimposed on an older one.*

if in its lifespan you neither breathe nor blink. So we, whose lives last only a few tens of years, are blind to changes taking centuries, or even hundreds of millions of years. But tombstones become covered with lichen; wind and rain, frost, and the chemical action of the atmosphere (which rusts iron) destroy the stone itself. Below every cliff, pieces of broken rock,

*Evidence of destruction, Mojave Desert, California. Weather breaks the black rock (basalt) into pieces, which fall to lower levels and accumulate in a small cone.*

brought down by gravity, accumulate as scree, silent witness to unrelenting destruction. A waterfall wears away the rock; and migrates upstream, leaving a gorge to mark its passage. Hutton, who was a farmer, recognised that soil, essential to the growth of plants and the food supply of animals, would not exist without the 'necessary decay' of rocks by weathering and erosion. Long continued, the process would, however, ultimately destroy all land.

The sea, too, destroys the land. Like a horizontal saw, it undercuts the cliffs and works to reduce the land to sea-level.

*Erosion by the sea, Pacific coast, Baja California. Like a horizontal saw, the sea undercuts the cliffs, which retreat, leaving a wave-cut bench behind.*

Hutton observed that round our coasts wave-cut benches are often found above reach of the sea, and concluded that these raised beaches prove that the sea once stood higher relative to the land. Like tree rings, raised beaches at different levels record past events. Perhaps the land can rise and be restored.

Just as we do, rocks respond to environmental change – though much more slowly. Like us, they are stable only within limited ranges of temperature and pressure. Under the 'right' conditions, minerals exchange material with their environment. This is called 'metamorphism': we do the same and call it

*Raised beach, Isle of Arran. The wave-cut platform is evidence that the sea formerly stood at a higher level relative to the land. Hutton observed raised beaches around Scotland, and correctly understood their significance.*

feeding. Rocks that have been formed deep within the Earth decompose when exposed at the surface, while those formed at the surface are re-crystallised and transformed by metamorphism when deeply buried. When the environment is no longer favourable to us, we change too: 'Earth to earth, ashes to ashes, dust to dust'.

Instead of looking in written records, James Hutton consulted the testimony of the rocks. He believed that our ability to re-create the past depends upon adequate understanding of the processes now operating. As Sir Archibald Geikie put it: *The Present is the key to the Past.* We may, of course, be overlooking events too rare to have been witnessed in recorded history. The collision of a comet with Jupiter in 1994, for example, raised interest in possible collisions with the Earth. But Hutton saw that slow-acting processes can bring about great changes — provided there is enough time.

Hutton's detective work proved that the Earth was far older than had been thought, and opened the way that led to our present understanding of Earth history. To arrive at an absolute age requires knowledge of the rate of decay of naturally occurring radioactive elements, such as uranium; but radioactivity was unknown until 100 years after Hutton's death. Nevertheless, what Sir Isaac Newton did for our understanding of space, Hutton achieved a century later for our appreciation of the immensity of time.

> 'The shepherd', Hutton wrote, 'thinks the mountain, on which he feeds his flock, to have always been there. But the man of scientific observation, who looks upon the chain of physical events connected with the present state of things, sees great changes that have been made, and foresees a different state that must follow in time, from the continued operation of that which actually is in nature ... [Go to the mountains] to read the immeasurable course of time that must have flowed during those amazing operations which the vulgar do not see, and which the learned seem to see without wonder!'

*River erosion, Grand Canyon of the Yellowstone River. Because the waterfall continues to be undercut, its position slowly moves upstream. The canyon marks the route that has been taken by the receding fall.*

CHAPTER TWO

# FIELD OBSERVATIONS

JAMES HUTTON was born in Edinburgh on 3 June 1726. He entered Edinburgh University at the age of fourteen, with an early interest in chemistry. Ever practical, he and his friend James Davie extracted sal ammoniac ($NH_4Cl$) from the soot of Edinburgh's chimneys. This salt was used for dyeing and working with brass and tin, and so their pioneering work in industrial chemistry became a profitable business. Medicine being closely linked to chemistry, Hutton enrolled as a medical student in Edinburgh (1744–47). After studying in Paris, he obtained an MD degree at Leyden in 1749 with a thesis titled *On Circulation of Blood in the Microcosm*. The title reminds us that scientists and philosophers were often influenced by analogies between the human body (the *microcosm*) and the world at large (the *macrocosm*). Indeed, nearly forty years later, one of Hutton's great contributions to science was his demonstration of circulation in the physical world: new rocks are made out of the debris of older rocks.

Returning to Scotland in 1750, Hutton abandoned medicine, and moved instead to a Berwickshire farm, 50 miles (80 km) from Edinburgh, which he had inherited from his father. Determined to study agriculture as a scientific discipline, he went to Norfolk in 1752 to learn innovative farming. From there he travelled on foot throughout England. On those expeditions he began to study geology, looking with curiosity into every ditch and river bed he found. He also travelled through Holland, Belgium and France, noting geological features.

Hutton returned to Scotland in 1754, and for thirteen years concentrated on improving his farm. His interest in geology continued, and in 1764 he made a tour to the north of Scotland, passing through Crieff, Dalwhinnie and Fort Augustus to Inverness, then north to Caithness, before returning by Aberdeen.

In 1767 Hutton moved permanently to Edinburgh, where he took an active part in the brilliant society that was the focus of the Scottish Enlightenment. Among the luminaries involved were David Hume, Adam Smith, Joseph Black, John Clerk of Eldin, Robert Adam, Adam Ferguson and William Robertson. A younger generation included John Playfair, John Clerk junior and Sir James Hall of Dunglass.

Hutton's travels always had a geological purpose. For about thirty years, as Playfair reports, Hutton 'never ceased to study the natural history of the globe, with a view of ascertaining the changes that have taken place on its surface, and of discovering the causes by which they have been produced'. While searching for granite, Hutton tells us, he 'examined Scotland from the one end to the other [and] almost all of England and Wales, excepting Devon and Cornwall'. He 'travelled every road from the borders of Northumberland and Westmoreland to Edinburgh'. From there he 'travelled to Port-Patrick, and along the coast of Galloway and Ayrshire to Inverary in Argyleshire'. Whether on foot, on horseback, or by chaise, travelling was exhausting and painful, yet he had 'examined every spot between the Grampians and the Tweedale mountains from sea to sea, ... through Caithness to the Pentland-Frith or Orkney Islands'.

Hutton had a remarkable knowledge of rocks and their distribution. 'I think', he wrote, 'I can undertake to tell from whence had come a specimen of gravel taken up anywhere, at least upon the east side of this island. Nor will this appear in any way difficult, when it is considered, that, from Portland [on the English Channel] to the Orkneys, there are at least ten different productions of hard stone in the solid land … with all of which I am well acquainted.' His knowledge was detailed: he observed, for instance, the exact dip (or inclination) of the beds of coal around Newcastle. He also affirmed that, contrary to popular belief, Britain is not formed primarily of granite. 'I would say that there is scarcely one five-hundred part of Britain that has granite for its basis', and he made a remarkably accurate assessment of the Southern Uplands: 'I had only seen granite in three places of this whole southern region, viz., at Buncle Edge in Berwickshire on the east, at Loch Doon in Ayrshire on the west, and Mount Criffel near Dumfries on the south.' Hutton was also able to distinguish between the softer strata of the Midland Valley and the harder 'alpine schistus' of the Highlands and Southern Uplands, which gave rise to more rugged terrain. Although Hutton looked for the junctions between the softer and harder strata, the Midland Valley of Scotland is bounded by geological faults – great fractures in the Earth's crust, running NE–SW, along which the rocks have been moved from their original positions. These well-known faults are the Highland Boundary Fault and the Southern Uplands Boundary Fault. The position of the first is visible as a straight line of hills when looking north to the Highlands from Perth; the second appears as a similar line when looking south towards the Lammermuirs from Edinburgh.

Hutton's understanding of the geological framework of Britain enabled him to lay the foundations of the emerging science of geology. His *Theory of the Earth,* published first as a paper by the Royal Society of Edinburgh and later as a book in two volumes by a commercial publisher, told how he knew that the Earth was immensely old. This was his crowning achievement.

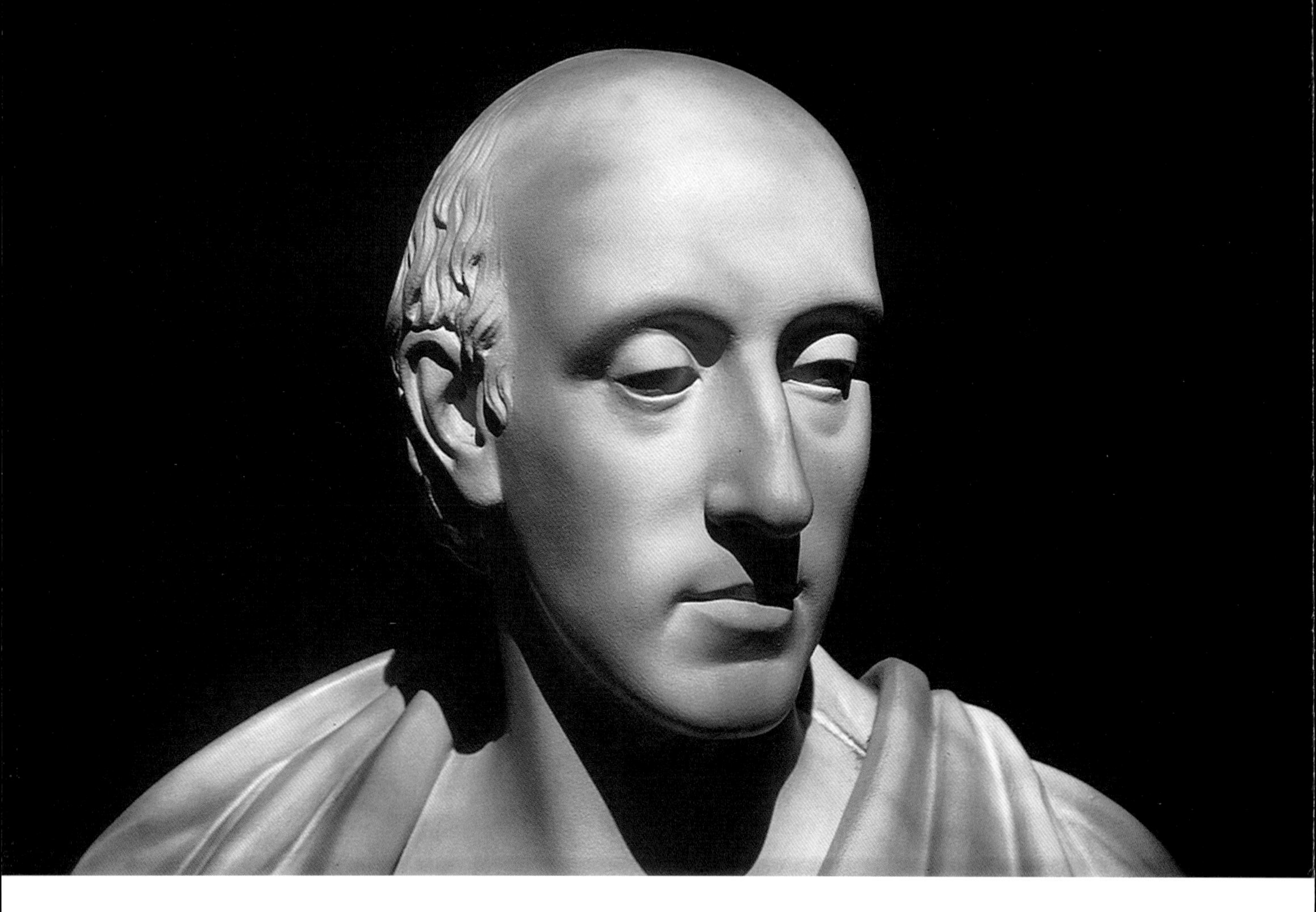

*James Hutton: Bust in the Royal Society of Edinburgh.*

CHAPTER THREE

# *PUBLICATION OF THE* THEORY OF THE EARTH

IN SIR HENRY RAEBURN'S portrait of Hutton, Robert Louis Stevenson saw 'Hutton the geologist, in quakerish raiment, looking altogether trim and narrow, as if he cared more about fossils than young ladies'. Hutton's letters, however, show his bawdy side and that he liked brandy toddy. He had an illegitimate son, although his friends did not know this until after his death, when his grandson asked John Clerk for help.

Playfair, Hutton's biographer, gives this word-portrait:

> His figure was slender, but indicated activity; while a thin countenance, high forehead, and a nose somewhat aquiline, bespoke extraordinary acuteness and vigour of mind. His eye was penetrating and keen, but full of gentleness and benignity.
>
> His conversation was inestimable; as great talents, the most perfect candour, and the utmost simplicity of character and manners, all united to stamp a value upon it. He had, indeed, that genuine simplicity, originating in the absence of all selfishness and vanity, by which a man loses sight of himself altogether, and neither conceals what is, nor affects what is not ... His conversation was extremely animated and forcible, and, whether serious or gay, full of ingenious and original observation. Great information, and an excellent memory, supplied an inexhaustible fund of illustration, always happily introduced, and in which, when the subject admitted of it, the witty and the ludicrous never failed to occupy a considerable place.
>
> A brighter tint of gaiety and cheerfulness spread itself over every countenance when the Doctor entered the room ... The acquisition of fortune, and the enjoyments which most directly address the senses, do not call up more lively expressions of joy in other men, than hearing of a new invention, or being

acquainted with a new truth, would, at any time, do in Dr Hutton. This sensibility to intellectual pleasure, was not confined to a few objects, nor to the sciences which he particularly cultivated; he would rejoice over Watt's improvements on the steam-engine, or Cook's discoveries in the South Sea, with all the warmth of a man who was to share in the honour or the profit about to accrue from them. The fire of his expression, on such occasions, and the animation of his countenance and manner, are not to be described; they were always seen with delight by those who could enter into his sentiments, and often with great astonishment by those who could not.

With this exquisite relish for whatever is beautiful and sublime in science, we may easily conceive what pleasure he derived from his own geological speculations. The novelty and grandeur of the objects offered by them to the imagination, the simple and uniform order given to the whole natural history of the earth, and, above all, the views opened of the wisdom that governs nature, are things to which hardly any man could be insensible; but to him they were matter, not of transient delight but of solid and permanent happiness.

He was, perhaps in the most enviable situation in which a man of science can be placed. He was in the midst of a literary society of men of the first abilities, to all of whom he was peculiarly acceptable ... He used also regularly to unbend himself with a few friends, in the little society known by the name of the Oyster Club. The original members of it were Adam Smith, Joseph Black, and Dr Hutton, and round them was soon formed a knot of those who knew how to value the familiar and social converse of these illustrious men. As all three possessed great talents, enlarged views, and extensive information, without any of the stateliness and formality which men of letters think it sometimes necessary to affect; as they were all three easily amused; were equally prepared to speak and to listen; and as the sincerity of their friendship had never been darkened by the least shade of envy; it would be hard to find an example, where everything favourable to good society was more perfectly united, and everything adverse more entirely excluded. The conversation was always free, often scientific, but never didactic or disputatious;

*Portrait of James Hutton by Sir Henry Raeburn. This important portrait was formerly considered to have been painted before the artist went to Rome in 1784. Dr David Mackie has, however, shown that it is much later; he dates it as c.1790 (Ph.D thesis, Edinburgh). Scottish National Portrait Gallery.*

> and as this club was much the resort of the strangers who visited Edinburgh, from any object connected with art or with science, it derived from thence an extraordinary degree of variety and interest.

When the Royal Society of Edinburgh was founded in 1783, Hutton, Black, Smith and Playfair were among the members of its first Council, and Hutton later joined Black as joint President of the Society's Physical Class. From 1786 until his last illness, Hutton frequently chaired meetings. When not presiding, he was often the speaker. His extraordinarily wide-ranging contributions included geology, medicine, agriculture, chemistry, meteorology, philosophy and the Chinese language.

According to Playfair, Hutton had outlined his theory about the age of the Earth several years before – sharing it only with Black and Clerk. 'He was in no haste to publish his theory; for he was one of those who are much more delighted with the contemplation of truth, than with the praise of having discovered it.' Despite initial reluctance, he agreed to read a paper to the Society, the topic to contain 'an examination of the System of the habitable Earth with regard to its duration and stability'. Both Hutton and Black were on the Publications Subcommittee, which may have provided stimulus, and the paper was presented at the March and April meetings, 1785. Hutton being indisposed, Black took his place at the March meeting. The Abstract of 1785 is the earliest published version of Hutton's theory, the full paper being included in the first volume of the Royal Society of Edinburgh's Transactions, published in 1788. Its title was 'Theory of the Earth; or an investigation of the laws observable in the composition, dissolution, and restoration of land upon the globe'.

Playfair thought, and later writers have agreed with him, that Hutton's style hindered the theory's acceptance: 'These defects produce a degree of obscurity astonishing to those who knew him, and who heard him every day converse with no less clearness and precision, than animation and force.' Hutton's supposed 'prolixity and obscurity' have kept many from reading what he wrote. Knowledge of his theory has usually been derived from Playfair's lucid *Illustrations of the Huttonian Theory*, a text (without graphic illustrations) published after Hutton's death to popularise the main features of the theory. Hutton's ideas are, however, easily enough understood by anyone reading his own words.

*Basalt dykes, Skelmorlie, on the Clyde. Seen and correctly interpreted by Hutton, these dykes were intruded as hot liquid (magma) into the surrounding conglomerates and sandstones.*

CHAPTER FOUR

# INTERPRETING THE TESTIMONY OF THE ROCKS

## James Hutton's *Theory of the Earth*

In his *Theory of the Earth*, Hutton described his observations and explained the far-reaching conclusions he drew from them. It is said that the best geologist is the one who has seen the most rocks, and it is doubtful if any contemporary knew the subject better than Hutton did. More than thirty years of field observation had convinced him that most rocks are the recycled products of older deposits. Stratified rocks are the commonest on the surface of the Earth, and he interpreted them as sediments derived by the weathering and erosion of an earlier land. Subsequently consolidated by heat, they were raised from the sea to form the present land. As the geological processes of erosion, sedimentation and uplift are very slow, the Earth must be vastly older than the age derived by literal interpretation of the Bible.

Hutton also challenged the accepted view that granite was the oldest rock of all, supposedly precipitated from a primeval ocean. He thought it more likely that, like basalt, it had formed from a molten state. In his *Theory of the Earth*, we see Hutton as the master of observation and deduction, guided always by the field evidence collected during his extensive travels. Begin, he insisted, by studying actual rocks and processes now operating: 'Let us, therefore, open the book of Nature, and read in her records.'

## Hutton's Specimens

Raeburn's portraits typically show objects illustrating the character of the sitter. The portrait of Hutton shows his quill pen, the massive manuscripts and some of the treasured specimens he enjoyed using to explain and illustrate his theory. 'God's books', he called them, 'wrote upon by Gods own finger.'

Playfair described Hutton's special skill as an observer:

> Long and continued practice had increased his powers of observation to a high degree of perfection … With an accurate eye for perceiving the character of natural objects, he had in equal perfection the power of interpreting their signification, and of decyphering those ancient hieroglyphics which record the revolution of the globe … There have been few who equalled him in reading the characters which tell not only what a fossil *is*, but what it *has been*, and declare the series of changes through which it has passed … None was more skilful in marking the gradations of nature, as she passes from one extreme to another; more diligent in observing the *continuity* of her proceedings, or more sagacious in tracing her footsteps, even where they were lightly impressed. [In the eighteenth century the word 'fossil' was used for any interesting object dug up from the ground.]
>
> With him [geology] was a sublime and important branch of physical science, which had for its object to unfold the connexion between the past, the present, and the future conditions of the globe. Accordingly his *collection of fossils* was formed for explaining the principles of geology, and for illustrating the changes which mineral substances have gone through.

Here, in his own words, we glimpse Hutton interpreting a specimen for a visitor: 'I shall only mention a specimen in my own collection. It is wood petrified with calcareous earth, and mineralized with pyrites. This specimen contains in itself, even without the stratum of stone in which it is embedded, the most perfect record of its genealogy. It had been eaten or perforated by those sea-worms which destroy the bottom of our ships [the shipworm *Teredo*]. There is the clearest evidence of this truth. Therefore, this wood had grown upon land which stood above the level of the sea, while the present land was only forming at the bottom of the ocean.'

*Sandstone blocks with water-rounded pebbles, Perth Bridge (1766–70), built by John Smeaton, engineer of the Forth and Clyde Canal. Hutton realised that many rocks, like these, had formerly been loose deposits transported by rivers towards the sea.*

*Pebbles sorted and rounded by wave-action, beach east of Hopeman, Banffshire coast. These pebbles are the tools the sea uses to attack the cliffs during storms.*

## Sedimentary Rocks

Rocks are destroyed by weathering and erosion, the debris being carried by rivers to the sea. In Hutton's words: 'From the top of the mountain to the shore of the sea, everything is in a state of change; the rock dissolving, breaking, and decomposing, *for the purpose* of becoming soil; the soil travelling along the surface of the earth, in its way to the shore; and the shore wearing and wasting by the agitation of the sea. Without those operations, which wear and waste the coast, there would not be wind and rain; and, without those operations which wear and waste the solid land, the surface of the earth would become sterile ... In the *necessary* waste of land which is inhabited, the foundation is laid for future continents, *in order to* support the system of this living world' (italics added).

Rocks, for Hutton, are in general strata of sandstone, gravel, conglomerate (pudding-stone), shale and limestone. Except for their compactness, strata of grit and sandstone are identical to the unconsolidated sand now forming on the shore. Sand is separated and sized by streams and currents; gravel is formed

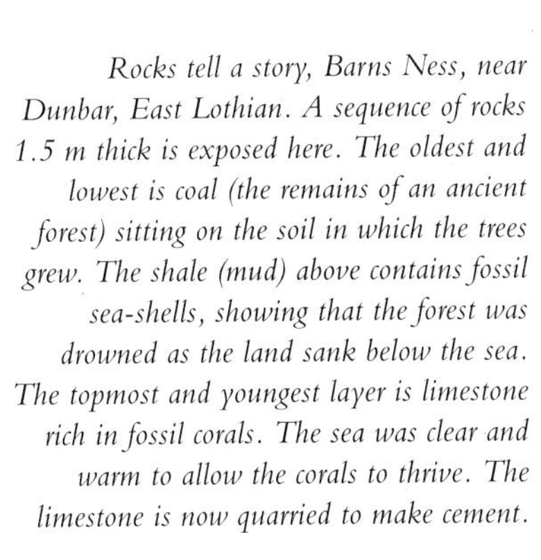

*Rocks tell a story, Barns Ness, near Dunbar, East Lothian. A sequence of rocks 1.5 m thick is exposed here. The oldest and lowest is coal (the remains of an ancient forest) sitting on the soil in which the trees grew. The shale (mud) above contains fossil sea-shells, showing that the forest was drowned as the land sank below the sea. The topmost and youngest layer is limestone rich in fossil corals. The sea was clear and warm to allow the corals to thrive. The limestone is now quarried to make cement.*

*Layers of sedimentary rock, Grand Canyon of the Colorado River. This thick pile of strata accumulated, layer upon layer, over a very long period of time, before being raised to its present level and cut through by the river.*

*The tools of the river. Colorado River at the bottom of the Grand anyon. These great boulders were used by the river to cut the canyon, and were rounded by attrition in the process. They are on the way downstream to the sea. The figure in blue gives the scale.*

by 'the attrition of stones agitated in water'; and shale accumulates by the sedimentation of mud. Hutton wrote:

> We are led to conclude that all the strata of the earth have had their origin at the bottom of the sea, by the collection of sand and gravel, of shells, and of earths and clays. Nine tenths, perhaps ninety-nine hundredths of this earth, *so far as we see*, have been formed by natural operations of the globe, in collecting loose materials, and depositing them at the bottom of the sea; consolidating those collections in various degrees, and either elevating those consolidated masses above the level on which they were formed, or lowering the level of that sea. The spoils or wreck of an older world are everywhere visible in the present. The strata which now compose our continents are all formed out of strata more ancient than themselves [italics added].

Pebbles within a conglomerate are sometimes themselves fragments of a still earlier conglomerate – evidence of more than one cycle of erosion and uplift. Hutton clearly grasped this point: 'The strata of the globe are also found composed of bodies which are fragments of former strata, which had already been consolidated, and afterwards were broken and worn by attrition, so as to be made gravel.'

*Conglomerate. These boulders, many of which are harder than steel, are almost as big as the water-rounded boulders in the Grand Canyon. The figure is the same one as in the previous picture, but the rock exposure is at Stonehaven, on the east coast just south of the Highland Boundary fault. The conglomerate (Hutton's 'pudding-stone') is evidence that a great river once came from a mountain chain much higher than the present Highland hills.*

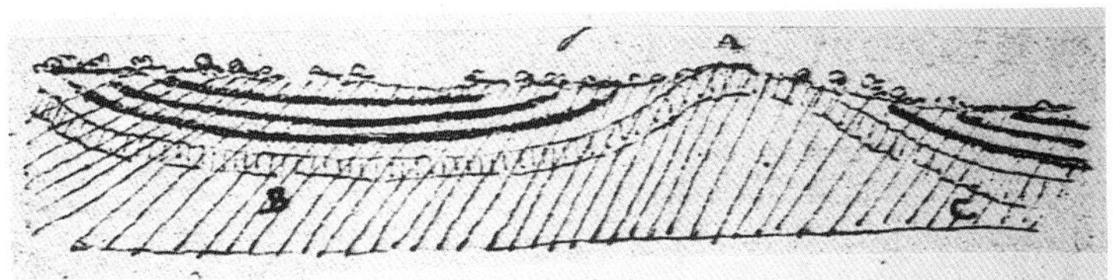


*Cross-section of the Midlothian Coal Basin, drawn by John Clerk of Eldin, accompanying Hutton in 1786. The strata, which include coal mined by the Clerk family and others, were originally deposited as horizontal layers, and later bent into arches (anticlines) and troughs (synclines) before being eroded to their present form. (Sir John Clerk of Penicuik.) Photograph by Gerhard Ott.*

## The Earth as a Heat Engine

Hutton thought the Earth was a heat engine - a device that turns thermal energy into mechanical energy - whose internal heat could convert loose material into hard rocks and raise them from the ocean floor to make new land. Most rocks are composed of silicates, which do not easily decompose when heated. Limestone, however, is a carbonate and behaves differently. He knew that limestone decomposes when heated, and the liberated carbon dioxide escapes into the atmosphere. He correctly surmised that, deep within the Earth, gas would not escape and, on cooling, limestone would re-crystallise as marble. Pressure does, in fact, influence chemical reactions.

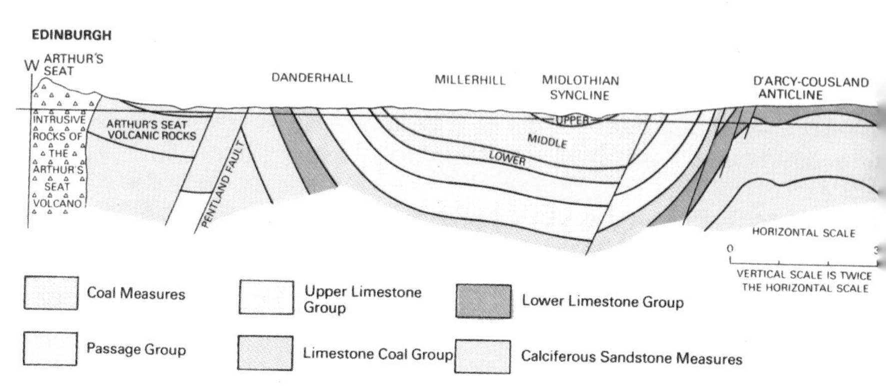


*Cross-section of the Midlothian Syncline. A modern cross-section for comparison with Clerk's section drawn in 1786,* British Regional Geology: the Midland Valley of Scotland, *1985, p. 132. (British Geological Survey.)*

Although strata forming at the bottom of the sea must be nearly horizontal, strata are actually found in every possible position from horizontal to vertical; individual layers are found bent and even doubled over. Hutton believed that the strata had been consolidated, uplifted and deformed, all by the power of the Earth's internal heat. Today we recognise that loose sand can be transformed into sandstone at low temperature when sand grains are cemented by minerals such as calcite ($CaCO_3$) deposited by percolating water. Hutton's term 'strata', however, included the layered 'greywackes' (alpine schistus) of the Southern Uplands and the crystalline schists, quartzites and marbles of the Highlands. In today's terminology all these are partially or wholly metamorphic. They have indeed been consolidated by heat and pressure.

Hutton was cautious about going beyond the evidence. Of the Earth's internal heat he asked: 'But how describe an operation which man cannot have any opportunity of perceiving? We only know, that the land is raised by a power

*Dipping sandstone strata, outcrop below the Castle at St Andrews. The bedding planes, which were originally horizontal, slope gently to the left. The steep edge on the right is a joint or fracture, often found in sedimentary rocks approximately perpendicular to the bedding layers.*

*Dip slope and escarpment, Kinnoull Hill looking east from Perth. This is the same structure as in the previous picture, but on a much larger scale. The layers (which in this case are ancient lava flows) dip or slope to the left (north) where they underlie (and are therefore older than) the red sandstones of Strathmore, which were quarried in Hutton's day to build Smeaton's Bridge at Perth.*

*Highly folded strata, River Avon, Tomintoul. Like other metamorphic rocks in the Highlands, these rocks have been squeezed by immense force and re-crystallised at high temperature.*

which has for principle subterraneous heat; but how that land is preserved in its elevated station, is a subject in which we have not even the means to form *conjecture*; at least we ought to be cautious how we indulge *conjecture* in a subject where *no means occur for trying* that which is but supposition' (italics added). Hutton's methodological point has been elaborated in our own day by the distinguished philosopher of science Sir Karl Popper in his *Conjectures and Refutations* (1989). Popper argued that knowledge advances by a succession of *conjectures*, each of which is tested and *refuted*, in the sense that an earlier conjecture is replaced by one that is more comprehensive. Others before Hutton had proposed geological theories, but he was the first to make systematic use of *refutation*. When Hutton was faced with deciding which of two competing theories is to be preferred – for example, whether granite is older or younger than the rocks that surround it – he predicted specific field observations that would be consistent with one theory and would refute the other, and then, with great skill, chose places where the predicted observations could be made.

Hutton wrote: 'We must not conclude that fire cannot burn in the mineral regions because our fires require the ventilation of the atmosphere; for much more powerful means *may* be employed by nature than those which we practise.' Although the word 'igneous' (like 'ignition') derives from the Latin word for fire, the 'fire' may be the glow of a hot body, as in an electric light. Such 'fire' is unlike a log fire, which involves chemical combination with atmospheric oxygen. Hutton's use of 'fire' may seem old-fashioned, but in January 1997 the technical journal *Science* reported: 'Since the 1960s, geophysicists have known that Earth's *internal fires* – stoked by heat leaking from the core and from radioactive decay in mantle rock – drive slow convection in the mantle, as a stove burner roils [stirs up] a pot of water' (italics added). This convection causes continents to move across the surface of the Earth, and is the basis of one of the most important discoveries of the twentieth century – plate tectonics.

## Igneous Intrusions

When sufficiently heated, rocks melt. Under pressure, molten rock (called 'magma') can be forced into fissures in other, still solid rocks. New rock, formed when magma solidifies on cooling below ground, is called an 'igneous intrusion', whereas magma reaching the surface in a volcanic eruption is called a lava flow. For Hutton, igneous intrusions were evidence of Earth's internal heat. He knew that throughout the Midland Valley of Scotland immense quantities of whinstone (basalt) have been forced to flow in a melted state, analogous to lava, among the sedimentary strata. Hutton was a pioneer in recognising igneous intrusions, but some of those bodies, as on Arthur's Seat and the Calton Hill in Edinburgh, are now known to be lavas rather than intrusions.

*Dyke near Crieff. Identified by Hutton in 1764, this dyke stands up like a wall because it is more resistant to weathering than the surrounding sandstones.*

*Dyke, Campsie Linn, north of Perth. This great dyke, which makes a waterfall in the River Tay, was recognised by Hutton, drawn by John Clerk of Eldin, and described by Sir Walter Scott in* The Fair Maid of Perth. *Hutton thought it might be an extension of the dyke at Crieff.*

Intrusions cutting across the strata are called 'dykes' or 'veins'; when injected parallel to the layers they are called 'sills'. On the road to Crieff in 1764, he saw a basalt dyke, which he thought might be the same as the one exposed at Dalcrue on the River Almond and at Campsie Linn on the River Tay, 'so it may be considered as having been traced for twenty or thirty miles, and perhaps much farther'. On his own doorstep in St John's Hill, Edinburgh, he recognised the Salisbury Crags as a sill. He had seen dykes at Skelmorlie on the Clyde, and was familiar with dykes exposed in the Water of Leith and in trenches in the New Town. The Castle Rock in Edinburgh is a cylindrical pipe that probably fed a volcano, long eroded away. Hutton had indeed ample evidence that sedimentary strata have been injected by hot molten rock (magma) from below.

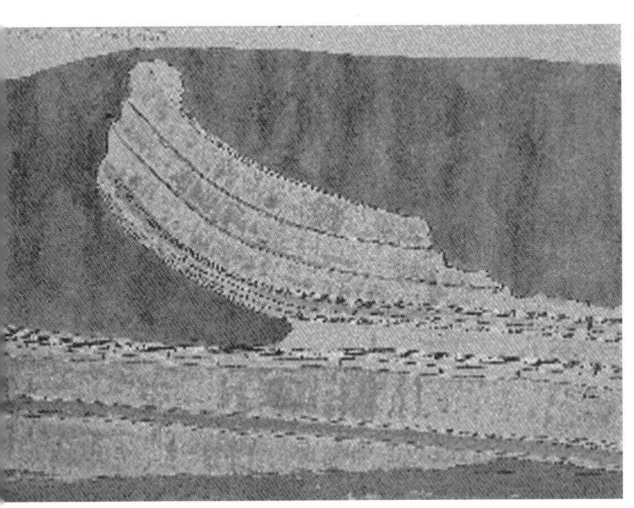

*Base of Salisbury Crags, Edinburgh, by John Clerk of Eldin. Part of an extensive drawing of the base of the intrusive sill. (Sir John Clerk of Penicuik.)*

*This similar view shows the intrusive contact at the base of the sill. The red rock (basalt) cuts across the underlying sedimentary strata, partly engulfing a block.*

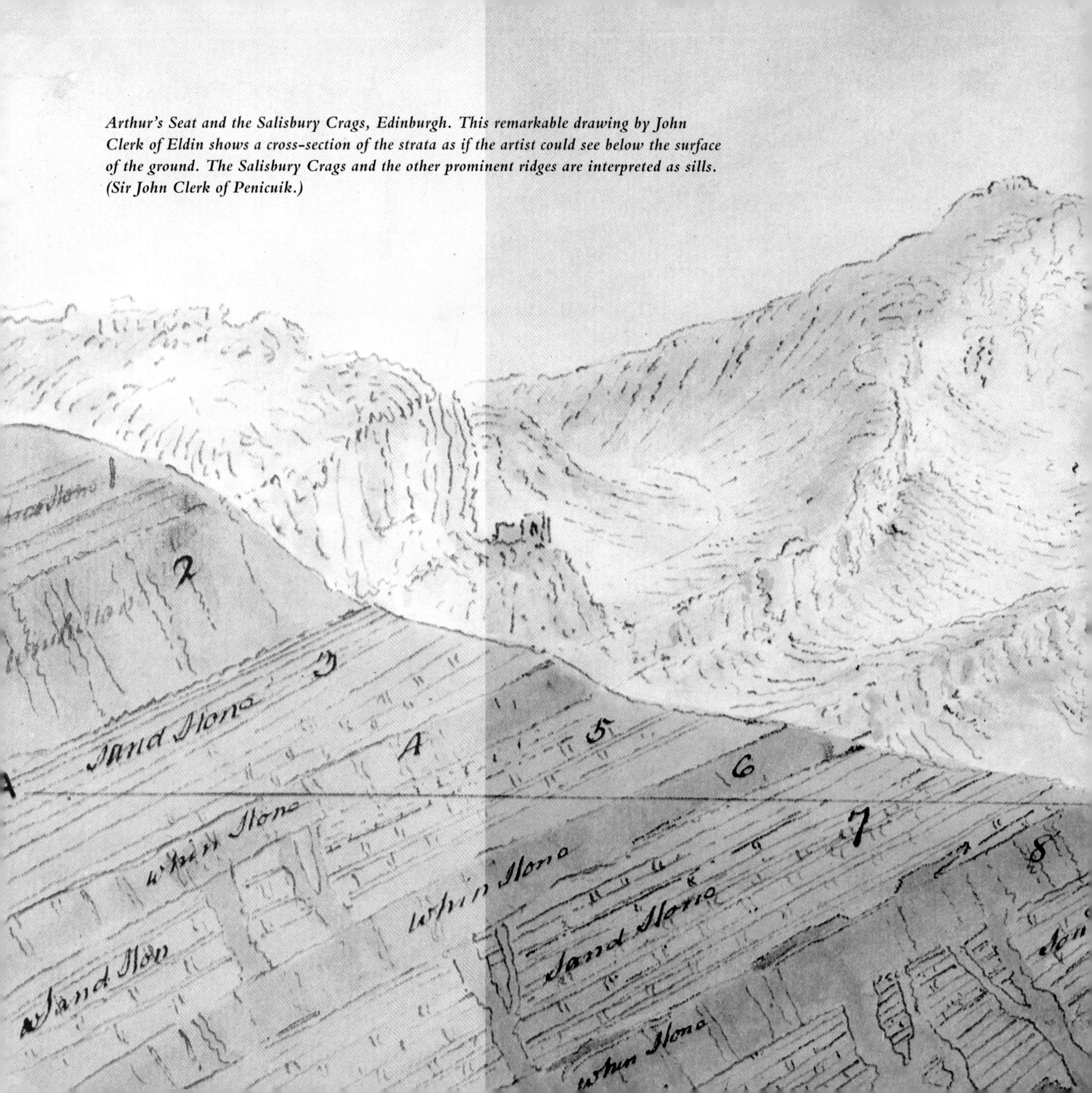


*Arthur's Seat and the Salisbury Crags, Edinburgh. This remarkable drawing by John Clerk of Eldin shows a cross-section of the strata as if the artist could see below the surface of the ground. The Salisbury Crags and the other prominent ridges are interpreted as sills. (Sir John Clerk of Penicuik.)*

*Detail from a similar view drawn by Robert Adam. Pen and brown ink, watercolour. This picture is perhaps the finest of over 1000 of Adam's picturesque drawings and watercolours. It illustrates the influence of the great architect on the artistic style of his brother-in-law, John Clerk of Eldin. If the landscape is Arthur's Seat and the Salisbury Crags, the castle takes the approximate place of Holyrood Palace. Notice also the small ruin of St Anthony's Chapel in John Clerk's picture. (From a private collection.) Photograph by Gerhard Ott.*

## Granite

Before writing his *Theory of the Earth*, Hutton had only seen granite outcrop in Aberdeenshire, where the evidence as to its origin was inconclusive. 'We shall, therefore,' he wrote, 'only now consider one particular species of granite; and if this shall appear to have been in a fluid state of fusion, we may be allowed to extend this property to all the kind.' His specimen was from near Portsoy, on the Banffshire coast. 'I have not been on the spot,' he reported, 'but am informed that this rock is immediately connected with the common granite of the country.' He described and illustrated an intergrowth of quartz and feldspar, still called 'graphic granite' from its supposed resemblance to Hebrew script.

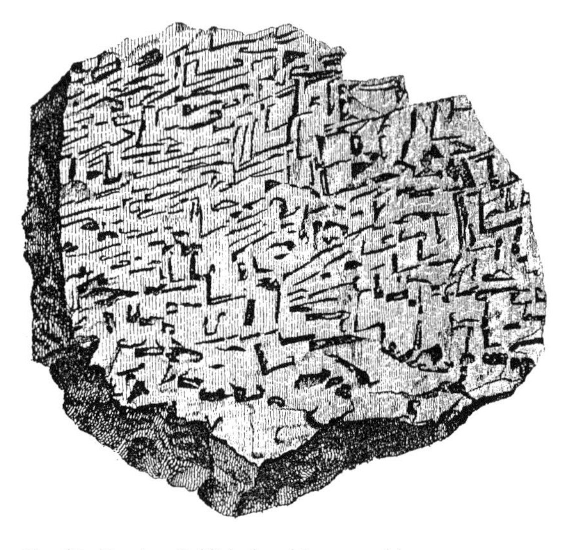

*Graphic Granite. Published and interpreted by Hutton in his* Theory of the Earth*, 1788.*

The specimen was probably provided by John Clerk, who visited Portsoy in 1779. On the way, Clerk had seen a granite vein cutting across the strata in the River Garry, and recorded that he 'took a specimen'. He found more granite veins at Portsoy, where he observed that the sea stacks 'shoot up like sal ammoniac' – an analogy likely to occur only to someone familiar with sal ammoniac production in Hutton and Davie's chemical works in Edinburgh.

Hutton concluded: 'It is not possible to conceive any other way in which those two substances, quartz and feldspar, could be thus concreted, except by congelation [crystallisation] from a fluid state, in which they had been mixed. Upon the whole, therefore, there is sufficient evidence of this body having been consolidated by means of fusion, and in no other manner.' It was not, in fact, until 1986 that the curious texture was reproduced artificially by simultaneous crystallisation of quartz and feldspar from a melt.

Hutton was 'particularly anxious about this subject of granite'. Other naturalists believed that granite was the oldest of all rocks, formed under conditions different from any we know today. Hutton, on the other hand, was sure that granite

*Sea stack at Portsoy. In 1779 John Clerk of Eldin described these stacks as 'shooting up like sal ammoniac' – a clear reference to Davie and Hutton's industrial process.*

was formed 'according to the laws of nature which we may investigate'. The Portsoy specimen persuaded Hutton that granite was probably an igneous rock, formed by melting stratified rocks. He thought that perhaps granite 'should be considered as a mass of subterraneous lava, which had been made to flow in the manner of our whinstone or basaltes'. This conjecture was directly opposed to the commonly held view that granite was the oldest of all rocks.

Hutton claimed that the sedimentary origin of so many rocks is evidence of a 'succession of worlds' – a system apparently designed to maintain a habitable Earth. It is to this that Hutton refers in the final sentence of his 1788 paper: 'The result, therefore, of our present enquiry is, that we *find* no vestige of a beginning, – no prospect of an end' (italics added). His opponents distorted this statement, pretending that Hutton claimed (like the Steady State Theory of modern cosmology) that there had been no beginning and would be no end. He was accused of atheism.

As there were two conflicting conjectures about the origin of granite, it was necessary to devise a test that would refute one but not the other. For this purpose Hutton set out to find key localities where the necessary evidence was likely to be found. In making these excursions around Scotland, Hutton was often accompanied by his devoted friend and 'coadjutor', John Clerk of Eldin, who not only helped as geological scout, but used his considerable artistic skill to make valuable drawings and field sketches of their discoveries.

*Sublimated sal ammoniac. A modern, artificial specimen for comparison with the Portsoy sea stack.*

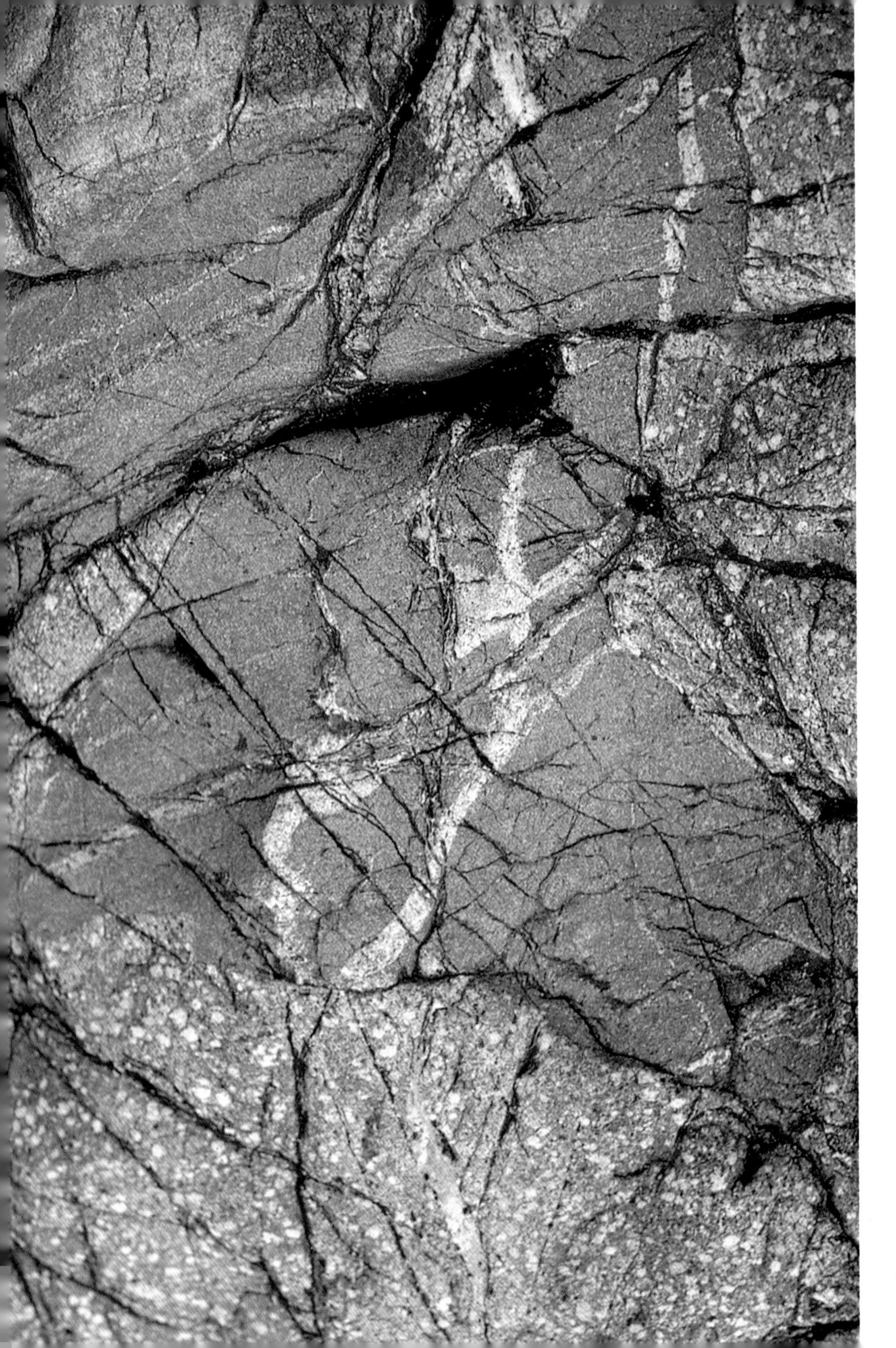

*Veins showing the intrusive nature of granite, Glen Tilt. This is one of the outcrops used by Hutton to demonstrate the igneous origin of granite.*

CHAPTER FIVE

# CONFIRMATION BY FIELD EVIDENCE

IMMEDIATELY after oral presentation of his theory in 1785, Hutton set out to put his theory to the test. Whereas ideas in physics and chemistry are tested by experiments, Hutton showed that geological conjectures can be tested by further field observations. Editing the third volume of Hutton's *Theory of the Earth* in 1899, Sir Archibald Geikie wrote: 'All his journeys in Scotland in quest of granite junctions, beginning with the visit to Glen Tilt where, for the first time, he saw intrusive veins and dykes proceeding from a mass of granite, were made for the purpose of proving the truth of a theory which he had already propounded and published.'

*Further examples of Granite veins, Glen Tilt.*

## Glen Tilt, 1785

If granite is the oldest rock, fragments of it might be found enclosed in *younger* rocks. But, as we have seen, on the basis of a specimen he had examined Hutton believed that granite had been intruded in a molten state into *older* rocks, and if this is so, some confirmation must appear at those places where the granite and the schistus are in contact. One might expect veins of the granite, which had been molten, to penetrate into the schistus, which had been solid. Hutton was anxious that 'an *instantia crucis* might subject his theory to the severest test'. A 'crucial instance' (Francis Bacon's term) is an example that decides between two competing conjectures. Hutton's methodology was impeccable.

Hutton inquired: 'Where might it be likely to find the junction of the granite with the strata?' Knowing there was granite in the sources of the River Dee, and schist in the sources of the River Tay, the country between held promise. Having seen 'abundant gravel formed of granite in the bed of the Tay', he was confident in finding the junction in its north-eastern tributaries. Because Clerk had already seen granite veins in the River Garry, Hutton and Clerk went to Glen Tilt (a tributary of the Garry), 'to visit the mountains from whence the granite rolling in the river came'. They found 'the granite breaking and displacing the strata in every conceivable manner, including fragments of broken strata, and interjected in every possible direction among the strata'.

Playfair said: 'The sight of objects which verified at once so many important conclusions in Dr Hutton's system, filled him with delight; and as his feelings, on such occasions, were always strongly expressed, the guides who accompanied him were convinced that it must be nothing less than the discovery of a vein of silver or gold, that could call forth such strong marks of joy and exultation.' Hutton's enthusiasm is striking: 'In matters of science curiosity gratified begets not indolence but new desires ... In the valley of Tarf [a tributary of the Tilt] we had the satisfaction to find many tumbled stones [water-rounded boulders], composed of broken schistus including granite. One of these pieces of stone had also this particular, of containing a vein which traversed both the mass of granite and broken schistus; consequently here is the proof of another operation subsequent to the fracture of this schistus and the injection of the granite.' Clerk's original drawing of that boulder and a sketch map of glens Tilt and Tarf are preserved on the back of his watercolour of the old Bridge of Tilt.

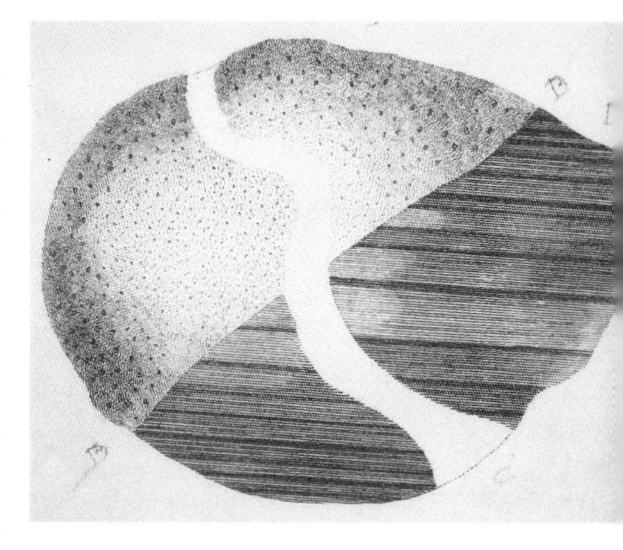

*Boulder with evidence of three events. An engraving based on a drawing by John Clerk of Eldin (1785). Layered schist has been cut by granite, both being cut by a still younger vein of 'red porphyry'. The engraving was intended to illustrate the third volume of Hutton's* Theory of the Earth. *The Tarf is a tributary of the Tilt. (Sir John Clerk of Penicuik.)*

Where bicycle tracks cross on top of footprints, we know the tracks are younger than the footprints: *superposition* is evidence of a time-sequence. The granite is younger than the

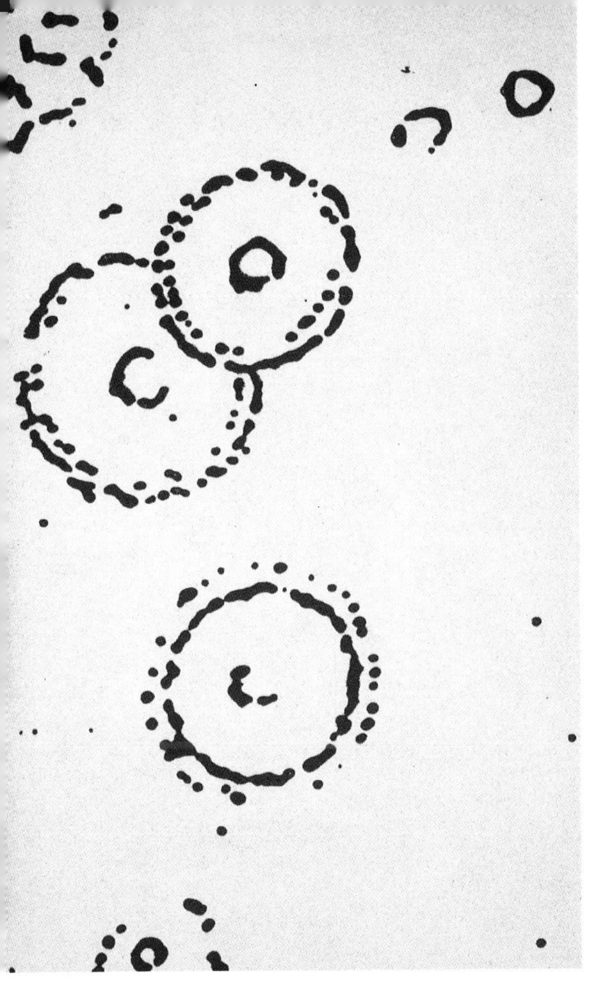

*Superposed lunar craters, drawn by John Clerk of Eldin, 14 May 1785. A young crater (probably Fracastorius) is superimposed on an older crater (probably Cyrillus). (Sir John Clerk of Penicuik.) Photograph by Jim Allan.*

schist it cuts; the vein, which cuts both, is younger still. Only four months before the visit to Glen Tilt, Clerk had been observing craters on the Moon. He showed a clear instance of a younger crater superimposed on an older one — just like pits in mud made by rain drops — a later pit partly obscuring an earlier one. Knowing, as they did, that the Glen Tarf boulder records a sequence of events, Clerk and Hutton must surely have seen that they could read the Moon's history as well as the Earth's.

## Galloway, 1786

The following year Hutton and Clerk set off to look for 'something decisive with regard to granite' in the Southern Uplands. On the way they travelled the Clyde coast, reporting on intersecting dykes (once again using superposition to determine age relationships), raised beaches (showing that sea-level had once stood higher relative to the land), and the general geological structure. Using granite boulders as clues, they made for Cairnsmore of Fleet, where they found that 'granite was the invading and schistus the invaded body ... We read the book so plainly open before us.'

Here is Hutton, the enthusiastic sixty-year-old field geologist at Sandyhills Bay on the Solway:

> The road was this way nearer, easier, and far more expeditious; but this was not our object; for now the rocky shore appeared, and we had every reason to expect to find something interesting in this critical spot. We therefore left the chaise, which we had for a long way attended on foot, to find its way up the hill, while we ran with some impatience along the bottom of the sandy bay to the rocky shore which is washed by the sea. We saw the schistus pretty erect, but variously inflected upon our right, where the land terminated in the sea. Upon the left we had the granite appearing through the sandy shore; and above, the granite hill seemed to impend upon the erected strata. We saw the place

*Vein of granite, Rockville, Colvend, Galloway. 'We found the granite ... interjected among the strata ... terminating in a thread where it could penetrate no farther' Hutton (1786).*

> nearly where the granite and the schistus must be united; but this place was bushy; and thus our fears and expectations remained for a moment in suspense. But breaking through the bushes and briars, and climbing up the rocky bank, if we did not see the apposition of the granite to the side of the erected strata so much as we would have wished, we saw something that was much more satisfactory, and to the purpose of our expedition. For here we found the granite interjected among the strata, in descending among them like a mineral vein, and terminating in a thread where it could penetrate no farther. Mr Clerk's drawing, and a specimen which I took of the schistus thus penetrated, will convince the most sceptical with regard to this doctrine of the transfusion of granite.
>
> We may now conclude, that, without seeing granite actually in a fluid state, we have every demonstration possible of this fact; that is to say, of granite having been forced to flow, in a state of fusion, among strata broken by a subterraneous force, and distorted in every manner and degree.

In other words, granite is demonstrably younger than the rocks it intrudes; consequently the conjecture that granite is a chemical precipitate from a primeval ocean is conclusively refuted.

## The Isle of Arran, 1787

Hutton set out with John Clerk's son, John (later Lord Eldin), and 'had but one object in view; this was the nature of the granite, and the connection of it with the contiguous strata'. Here again is the competent field geologist in action:

> By reason of moss and vegetation we had a very interrupted view of the immediate junction of the granite and schistus, which here appears in many places upon the summits of bare rock standing up among the heath and moss ... Having once got hold of the clue or caught the scent, we traced back (with more animation than could have been expected from such an innocent chase) the object of our investigation all the way to the Cataract rock. Great veins

*Granite vein, Colvend, Galloway, drawn by John Clerk of Eldin, accompanying Hutton in 1786. (Sir John Clerk of Penicuik.) Photograph by Gerhard Ott.*

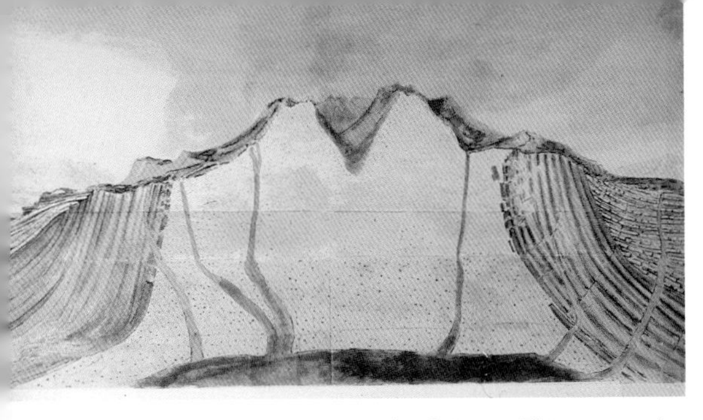

*Cross-section of the granite, Isle of Arran. This cross-section shows remarkable geological understanding. John Clerk junior (later Lord Eldin) accompanied Hutton to Arran in 1787 and made geological drawings. On some other occasion, John Clerk of Eldin also made drawings in Arran. We do not know for certain who drew this cross-section, but it was probably the father rather than the son. (Sir John Clerk of Penicuik.)*

*Cross-section of the granite, Isle of Arran. Sketch by John Playfair in a letter written to Sir James Hall, 10 October 1797, seven months after Hutton's death. 'The junctions I saw were I believe all visited by Dr H. At one of them I could see the marks of his hammer, (or at least I thought so), and could not without emotion think of the enthusiasm with which he must have viewed it. I was never more sensible of the truth of what I remember you said one day when we were looking at the Dykes in the Water of Leith since the Dr's Death, '"that these Phaeonomena [sic] had now lost half their value".' (GD 206/2/309, Dunglass Muniments, Scottish Record Office.) Photograph by Gerhard Ott.*

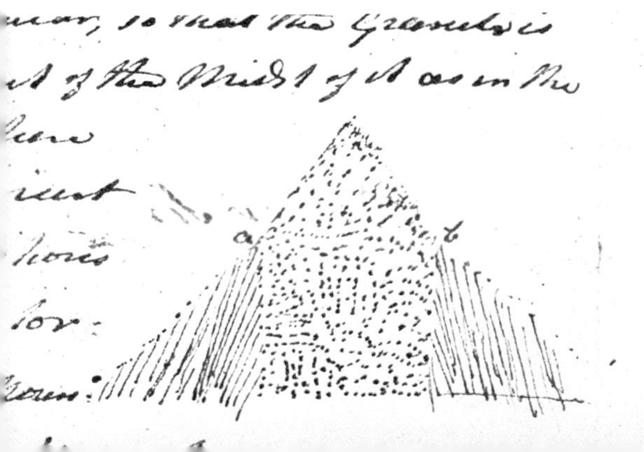

> of granite may be seen traversing the schistus, and ramifying in all directions. I procured a specimen, which I have had conveyed to Edinburgh, though weighing above 600 pounds.

In Arran, Hutton, at long last, found 'the immediate connection of the alpine schistus with the strata of the low country', but even today that particular exposure is difficult to interpret. Like all who have followed him to Arran, Hutton was impressed with 'the number and complexity of the dykes'. He observed composite intrusions, and recognised their age sequences. In a letter written shortly after Hutton had returned to Edinburgh, Sir James Hall wrote to his uncle: 'Dr Hutton was highly pleased with his expedition. He found dykes entirely formed of black glass, which you may suppose was a great treasure to him.' Glass was then being manufactured in Leith, and Hutton was familiar with the process: a mixture of sand and other substances is first melted and then cooled quickly. He correctly interpreted the dykes as igneous intrusions, comparing some of them to artificial samples found 'when green bottle glass is not perfectly vitrified by sufficient fusion, or has undergone some change in cooling'. The existence of dykes made of glass was overwhelming evidence of their igneous origin, and, by extension, of the igneous origin of dykes of basalt, too.

Hutton argued that the little island of Pladda, off Kildonan, had once been part of Arran:

> No proposition in natural history, concerning what is past is more certain. Pladda is the intermediate step by which we may remount to this view of high antiquity. We see the destruction of a high island in the formation of a low one; and we may perceive the future destruction, not of the little island only, but also of the continent itself, which is in time to disappear. Thus Pladda is to the island of Arran what Arran is to the island of Britain, and what the island of Britain is to the continent of Europe.

## Jedburgh, 1787

After his return from Arran, Hutton discovered by chance the junction between vertical greywackes (schistus) and overlying horizontal layers of red sandstone near Jedburgh. The alternative conjectures were: firstly, that the greywackes had been tilted into their vertical position underneath the overlying sandstones, or second, that they had been tilted and eroded before the sandstones were laid down on top of them. The second possibility was the more probable because a bed of conglomerate separates the two rocks, and the stones in it appear to have belonged to the older, vertical strata.

Hutton published John Clerk's drawing of the unconformity in 1788.

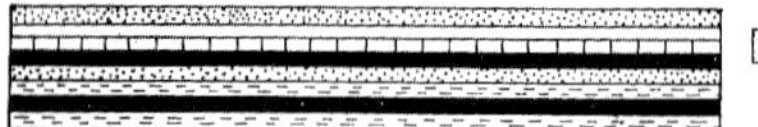

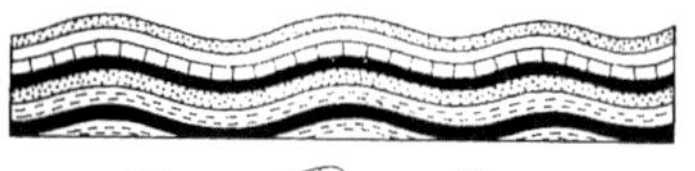

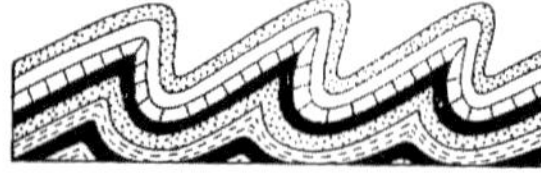

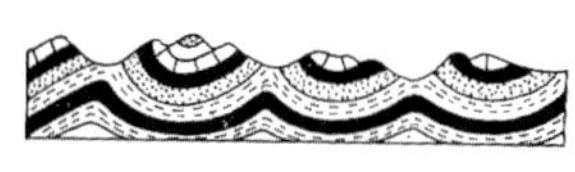

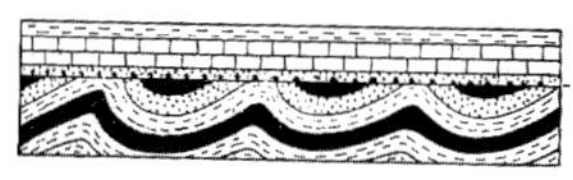


*The successive stages in forming an unconformity. Arthur Holmes,* Principles of Physical Geology.

## Siccar Point, 1788

The boundary between the resistant greywackes of the Southern Uplands and the softer sedimentary rocks of the Midland Valley runs through the Dunglass estate in East Lothian, 40 miles (65 km) east of Edinburgh. At Hutton's request Sir James Hall's uncle searched for and found the contact in the Tour (Tower) burn. Hutton and Playfair then joined Sir James Hall at Dunglass. Under the heading *The Theory confirmed from Observations made on purpose to elucidate the subject*, Hutton described what happened: 'Having taken boat at Dunglass burn, we set out to explore the coast. At Siccar Point, we found a beautiful picture of this junction washed bare by the sea. The sandstone strata are partly

*Horizontal strata, Jedburgh, south-east of Edinburgh. Red sandstones, deposited layer upon layer.*

*Folded strata near St Abb's Head, Berwickshire. Sir James Hall correctly interpreted these structures as resulting from horizontal compression of formerly horizontal strata.*

washed away, and partly remaining upon the ends of the vertical schistus; in many places, points of the schistus are seen standing up through among the sandstone, the greatest part of which is worn away. Behind this again we have a natural section of those sandstone strata, containing fragments of the schistus. Most of the fragments of the schistus have their angles sharp; consequently they have not travelled far, or been worn by attrition.'

Playfair's account is one of the classics of scientific literature:

> On us who saw these phenomena for the first time, the impression made will not easily be forgotten. The palpable evidence presented to us, of one of the most extraordinary and important facts in the natural history of the earth, gave a reality and substance to those theoretical speculations, which, however probable, had never till now been directly authenticated by the testimony of the senses. We often said to ourselves, What clearer evidence could we have had of the different formation of these rocks, and of the long interval which separated their formation, had we actually seen them emerging from the bosom of the deep? We felt ourselves necessarily carried back to the time when the schistus on which we stood was yet at the bottom of the sea, and when the sandstone before us was only beginning to be deposited, in the shape of sand or mud, from the waters of a superincumbent ocean. An epoch still more remote presented itself, when even the most ancient of these rocks, instead of standing upright in vertical beds, lay in horizontal planes at the bottom of the sea, and was not yet disturbed by that immeasurable force which has burst asunder the solid pavement of the globe. Revolutions still more remote appeared in the distance of this extraordinary perspective. The mind seemed to grow giddy by looking so far into the abyss of time; and while we listened with earnestness and admiration to the philosopher who was now unfolding to us the order and series of these wonderful events, we became sensible how much further reason may sometimes go than imagination may venture to follow.

The deformed and hardened rocks of the Southern Uplands and the Lake District were generally supposed to be relics of

*Unconformity at Siccar Point, Cockburnspath, East Lothian. The old strata were tilted on end and eroded before the younger sediments (made up of their debris) were deposited on the irregular surface. This famous exposure was predicted by Hutton, who was accompanied there by Playfair and Hall in 1788.*

*Siccar Point unconformity. The more resistant of the older strata project up into the younger, overlying rocks, showing that in former time conditions and processes were similar to those of today.*

the early crust of the Earth, but this conjecture was refuted when Hutton and Clerk found fossils in the Lake District (1788) and Sir James Hall found fossils on his way through the Southern Uplands from Edinburgh to Moffat (1792).

In a letter to James Watt in December 1791, Black reported that Hutton had been very ill and in great danger. By the time that letter was written, Hutton was much better, but he remained ill and in pain during his final years. Although it is unlikely he did any further fieldwork, he wrote a prodigious amount between 1791 and his death in 1797, including: *Dissertations ... in Natural Philosophy* (740 pages in length); ... *Principles of Knowledge* ... (2138 pages); ... *the Philosophy of Light, Heat, and Fire* (326 pages); the two-volume *Theory of the Earth* ... (1,187 pages) with additional chapters which remained in manuscript until 1899 (267 printed pages); the manuscript of his still unpublished *Elements of Agriculture* (1,045 pages); and various shorter contributions. Hutton is best remembered for his *Theory of the Earth*, in which he built on the material previously published by the Royal Society of Edinburgh and included the important evidence discovered on his excursions to Glen Tilt, Galloway, Arran, Jedburgh and Siccar Point.

*Sir James Hall's drawing of the Siccar Point unconformity. Sir James Hall*

*Edinburgh Castle looking south-west from John Playfair's monument on Calton Hill. In the foreground is the monument to Dugald Stewart (1753–1828), Professor of Mathematics and Moral Philosophy, and Adam Smith's biographer. To the right is the Scott monument on Princes Street; to the left is the North Bridge, linking the Old and New Towns. In front of the Bridge is the tower designed by Robert Adam for David Hume's grave; the tall obelisk commemorates the Martyrs of Parliamentary Reform, 1793–4. On the skyline, to the left of the obelisk, is the historic Kirk of St Giles on the High Street of the Old Town.*

CHAPTER SIX

# *HUTTON'S LEGACY*

EDINBURGH'S famous scenery expresses a rich geological history. Great variety in a small space is the result of an active past, and this variety provides an ideal laboratory for geological studies. During the Ice Age, a thick ice-sheet moved eastwards across the Edinburgh area. The Castle rock stood in its way, and the ice gouged the hollow of Princes Street Gardens and the Grass Market around it. Sandstones to the east were protected in the lee of the Castle rock and now form the ridge of the High Street descending from the Castle to the Palace of Holyrood. Beyond is Arthur's Seat, the relic of a tilted and deeply eroded volcano, of which Calton Hill is a fragment. The medieval city was confined to the ridge, on which tall tenements housed a compressed population. The Old Town is like herring bones, with the backbone of the High Street supporting closes and wynds at right angles. Sewage was thrown from the windows, and the High Street was an open drain. James Boswell, who escorted Dr Samuel Johnson in 1773, recorded Johnson's complaint: 'I smell you in the dark!'

*Cross-section of strata in Germany. Published in 1756 by J. G. Lehmann. An important early study of stratified rocks in the Harz Mountains, for comparison with John Clerk's detailed cross-section of a trench in Frederick Street. Lehmann recognised a sequence of rocks of different ages. Photograph by Gerhard Ott.*

When Castle Street and Frederick Street were developed in the New Town, Clerk seized the opportunity to record the rocks exposed during excavations. His technique of constructing squares on the walls of the trenches made it easier to 'map' the geological details, exactly as geologists do today when studying trenches across California's San Andreas fault. Two hundred years after Clerk, modern geologists use the same symbols to represent the different strata. To a geologist, Clerk's cross-sections look thoroughly up-to-date in contrast

*Trench across the San Andreas fault, Cajon Pass, California.*

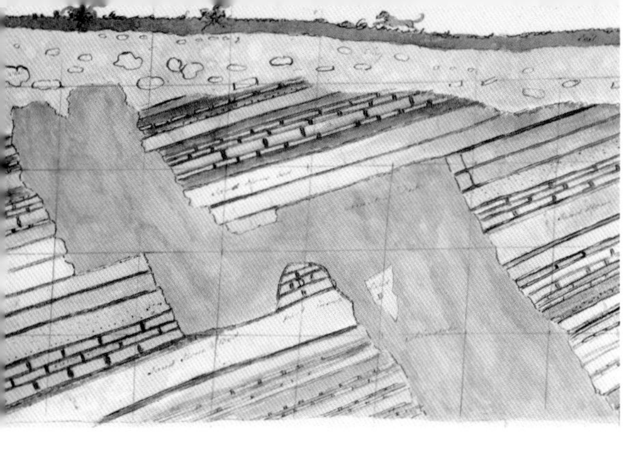

*East side of the drain in Frederick Street. This remarkable cross-section by John Clerk of Eldin was probably drawn in 1786 when the first feu was taken. A dyke cuts through the strata. The red squares were drawn to help the artist make an accurate record. Clerk's method and style are used today by engineering geologists studying trenches across the San Andreas Fault in California. (Sir John Clerk of Penicuik.)*

*Grid to aid survey of the trench across the San Andreas fault, Cajon Pass, California.*

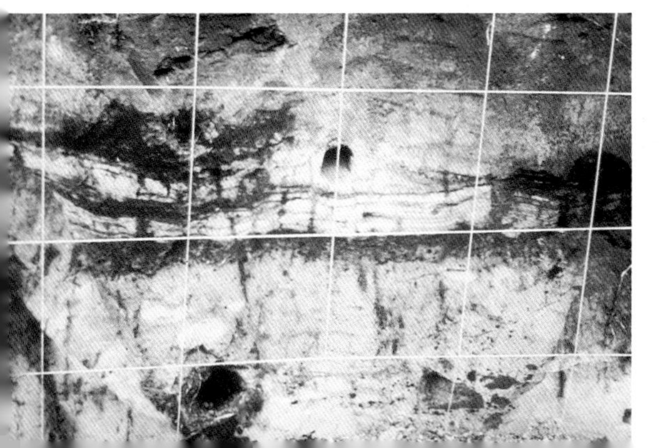

with sections drawn by other early geologists such as Nicholas Steno (1669) or J. G. Lehmann (1759).

## Hutton's Contributions to Geology

Although Hutton's greatest contribution was undoubtedly his recognition of 'deep time', a term made popular by John McPhee (1980), he also provided geologists with a valuable procedure – or, more precisely, demonstrated a method used in a less subtle way by miners. It does not matter whether a conjecture has been arrived at by observation, experience, or even by false analogy or theological supposition. What *does* matter is that we test it. To do this we must use the conjecture to make *verifiable* predictions, which is what Hutton did. Moreover, like a good detective on a case, Hutton displayed great skill in knowing what to look for and where to look for it. His specimens, for example, were not chosen for their superficial beauty. Each had been selected to illuminate an important principle, and Clerk's superb drawings supplemented the specimens by illustrating rock relationships on larger scales.

As an innovative farmer, Hutton was concerned about ecology, and as a medical doctor, he delighted in the study of the Earth's 'physiology'. He took a holistic view of this habitable Earth, recognising that it had risen from the wreck of a former world: 'Here is a compound system of things, forming together one whole living world.'

Most geologists learned about Hutton's theory from Playfair's fine prose, and this was so for Sir Charles Lyell. Lyell, born the year Hutton died, was the great populariser of geology in the nineteenth century. Eleven editions of his *Principles of Geology* were published in his lifetime. He was knighted, and when he died he was buried in Westminster Abbey. Hutton, in contrast, was buried in an obscure Edinburgh plot unmarked for 150 years. Lyell's system was based on Hutton's theory, though Lyell cited important research from a later time showing that successive groups of sedimentary strata contain distinct fossils. Darwin read the newly published first volume of Lyell's *Principles* (1830) on the voyage of the *Beagle*. Lyell had acknowledged his debt to Hutton by putting a quotation from Playfair's *Illustrations* before the title page.

Two more quotations from Playfair – this time paraphrasing Hutton's unpublished *Elements of Agriculture* – arrived on the *Beagle* with Lyell's second volume (1832): 'The inhabitants of the globe, like all other parts of it, are subject to change. It is not only the individual that perishes, but whole species.' These ideas were, of course, marvellously developed by Darwin in *The Origin of Species*.

If we lack knowledge of the vast extent of time, we are short-sighted and miss what otherwise would seem obvious. Lyell pointed out, for example, that if we were convinced that the great pyramid in Egypt had been raised in a single day, we would be justified in believing that it was the result of superhuman power. Nearly four centuries ago, Francis Bacon drew attention to the *conformable* shapes of the continents on either side of the Atlantic (1620). But contrary to what has sometimes been claimed, Bacon did *not* discover continental drift or modern plate tectonics. That would have been impossible for him. Without knowledge of deep time, the idea that the Atlantic had formed by separation of the continents was simply inconceivable. Hutton changed that.

The discovery of deep time was a necessary forerunner of Darwinian evolution. Indeed, as Sir Alister Hardy pointed out

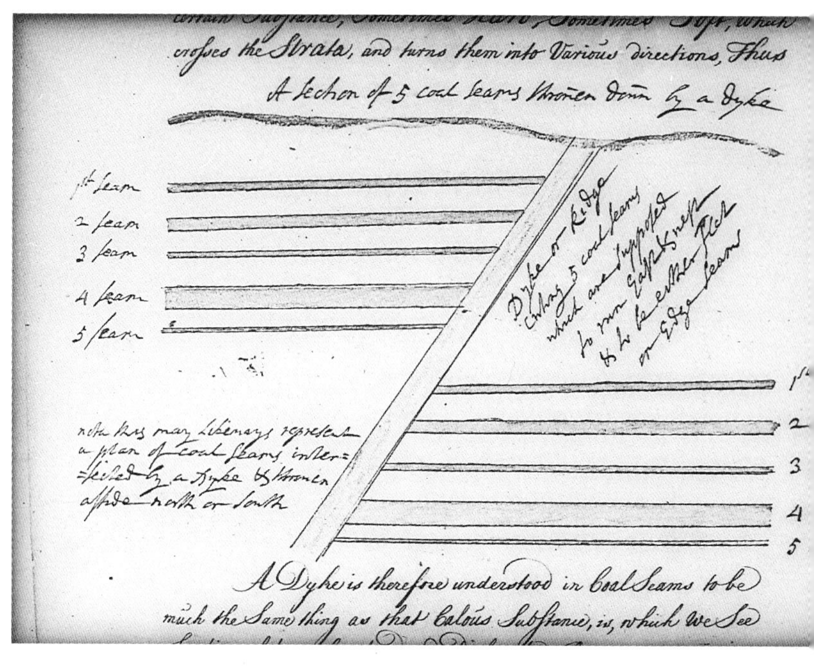


*Faulted coal seams. Can be considered either as a map or a cross-section.* Dissertation on Coal, *1740. Sir John Clerk, 2nd Baronet, father of John Clerk of Eldin. This illustration and the following one show that the Clerk family's practical knowledge of rock strata extended back a hundred years before Hutton's* Theory of the Earth *was published. (Sir John Clerk of Penicuik.) Photograph by Gerhard Ott.*

in his Gifford Lectures (1965), Hutton recognised natural selection as 'a beautiful contrivance'. In *The Origin of Species* (1859) Darwin wrote: 'It is hardly possible for me to recall to the reader who is not a practical geologist, the facts leading the mind feebly to comprehend the lapse of time. He who can read Sir Charles Lyell's grand work on the Principles of Geology, which the future historian will recognise as having produced a revolution in natural science, and yet does not admit how vast have been the past periods of time, may at once close this volume.' But Lyell and Darwin owed the concept to Hutton, who, before they were born, had stressed: 'What more can we require? Nothing but time.'

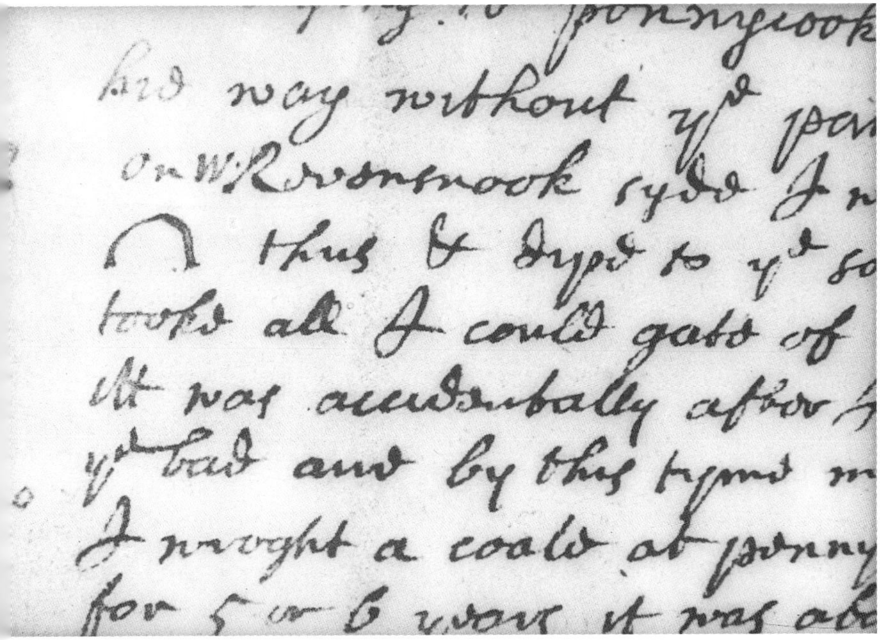

Hutton was right: 'How shall we acquire the knowledge of a system calculated for millions, not of years only, nor of the ages of man, but of the races of men, and the succession of empires? We must read the transactions of time past in the present state of natural bodies'. Today we can acknowledge the mute testimony of the rocks, knowing that the Present is the key to the Past.

*Anticlinal fold (arch). 'Advyse for my sone John anent working of coall, 12 Jan. [16]87.' Item 2 'On W. Ravensnook syde I wroght a coal about 4 foot thick it lay sadle ways ∩ thus & dyed to ye south'. Sir John Clerk, 1st Baronet, grandfather of John Clerk of Eldin. (Sir John Clerk of Penicuik.) Photograph by Gerhard Ott.*

In a classic textbook, Professor Arthur Holmes used Hutton's metaphor: 'To the geologist a rock is … a page of the Earth's autobiography with a story to unfold.' Hutton showed how to read it; doing so he disclosed the marvel of deep time. We have come to know that living creatures evolve, that continents drift, that even stars are born, mature, grow old and die. James Hutton opened our minds to these wondrous possibilities.

# *Further Reading*

Dean, D. R., *James Hutton and the History of Geology*, Cornell University Press, Ithaca, New York, 1992.

Hutton, J., *James Hutton in the Field and in the Study*, edited by Dennis R. Dean, Scholars' Facsimiles and Reprints, Delmar, New York, 1997.

## Also

Brush, S. G., *Transmuted Past: The Age of the Earth and the Evolution of the Elements*, Cambridge University Press, 1996.

Daiches, D., Jones P., and Jones, J., (eds), *The Scottish Enlightenment, 1730–1790: Hotbed of Genius*, Saltire Society, Edinburgh, 1996.

Dalrymple, G. B., *The Age of the Earth,* Stanford University Press, 1991.

McIntyre, D. B., *James Hutton and the Philosophy of Geology*, in, *The Fabric of Geology*, edited by Claude C. Albritton, Jr. Addison-Wesley, Reading, Massachusetts, 1963, for the 75th Anniversary of the Geological Society of America.

McIntyre, D. B., *James Hutton and his Edinburgh: The Historical, Social, and Political Background*. To be published in the *Proceedings of Hutton/ Lyell Conference*, 1997.

Toulmin, S., and Goodfield, J., *The Discovery of Time*, Harper & Row, 1966.

# Index

**Published by The Stationery Office and available from:**

**The Stationery Office Bookshops**
71 Lothian Road, Edinburgh EH3 9AZ
(counter service only)
49 High Holborn, London WC1V 6HB
(counter service and fax orders only)
Fax 0171-831 1326
68-69 Bull Street, Birmingham B4 6AD
0121-236 9696 Fax 0121-236 9699
33 Wine Street, Bristol BS1 2BQ
0117-926 4306 Fax 0117-929 4515
9-21 Princess Street, Manchester M60 8AS
0161-834 7201 Fax 0161-833 0634
16 Arthur Street, Belfast BT1 4GD
01232 238451 Fax 01232 235401
The Stationery Office Oriel Bookshop
The Friary, Cardiff CF1 4AA
01222 395548 Fax 01222 384347

**The Stationery Office publications are also available from:**

**The Publications Centre**
(mail, telephone and fax orders only)
PO Box 276, London SW8 5DT
General enquiries 0171-873 0011
Telephone orders 0171-873 9090
Fax orders 0171-873 8200

**Accredited Agents**
(see Yellow Pages)

*and through good booksellers*

Printed in Scotland for The Stationery Office Limited by CC No. 10251 15C 7/97